COUNTDOWN TO SPACE

THE COLUMBIA SPACE SHUTTLE DISASTER

From First Liftoff to Tragic Final Flight

Michael D. Cole

Enslow Publishers, Inc.

40 Industrial Road   PO Box 38
Box 398   Aldershot
Berkeley Heights, NJ 07922   Hants GU12 6BP
USA   UK

http://www.enslow.com

Revised edition of *Columbia: First Flight of the Space Shuttle* © 1995

Library of Congress Cataloging-in-Publication Data

Cole, Michael D.
    The Columbia space shuttle disaster: from first liftoff to tragic final flight /
Michael D. Cole.
        p. cm. — (Countdown to space)
    Rev. ed. of: Columbia. © 1995.
    Summary: Details the first flight of the space shuttle Columbia, as well as its
tragic final flight.
    Includes bibliographical references and index.
    ISBN 0-7660-2295-1 (hardcover)
    1. Columbia (Spacecraft)—Juvenile literature. 2. Space vehicles—Accidents—
Juvenile literature. [1. Columbia (Spacecraft) 2. Columbia (Spacecraft)—
Accidents. 3. Space shuttles.] I. Title. II. Series.
TL795.5.C6523 2003
629.44'1—dc21

                    2003004823

Printed in the U.S.A.

10 9 8 7 6 5 4 3 2 1

CONTENTS

The space shuttle Columbia waiting on the launchpad at the Kennedy Space Center in Florida.

1

A New Kind of Spaceship

The year was 1981. It had been six years since the last U.S. astronaut had rocketed into space. On that last flight, astronauts had linked up in space with Soviet cosmonauts for the Apollo-Soyuz mission.

The National Aeronautics and Space Administration (NASA) was about to begin a new era in space travel. It took more than ten years to develop, but the first space shuttle, *Columbia*, now sat on the launchpad at the Kennedy Space Center in Florida.

But it had been sitting there for three months.

Technical problems with the shuttle had delayed the launch several times. It was an incredibly complex machine.

When was it finally going to fly into space? Would it ever fly at all? No one knew for sure.

Columbia was a very complex machine. It would be the first reusable spaceship.

The first flight of the space shuttle *Columbia* would be a test flight different from any other in the history of the space program. Always in the past, when a new spaceship was first launched, it was an unmanned flight controlled from the ground. But the space shuttle was not like any other spacecraft.

The space shuttle would be the first *reusable* spaceship. It would lift off from a launchpad like previous spacecrafts, but it would not splash down in the water at the end of its flight. It would land on the ground like an airplane. To do that, it needed a pilot.

On *Columbia*'s very first flight, astronauts would be aboard. John Young and Robert Crippen were chosen for this challenging and dangerous task. Young was the most experienced astronaut in the program. This would be his fifth space flight. Crippen would be making his first flight into space.

Now, after three months of delays, NASA was ready

to try a launch on April 12, 1981. John Young and Robert Crippen woke before dawn. They ate the traditional NASA prelaunch breakfast of steak and eggs. Soon they were suiting up in their brown spacesuits and white helmets. Young and Crippen tested the suits, then removed the helmets. A van then carried the two astronauts three miles to the launchpad where the mammoth spaceship awaited them.

The pad was quite a sight. There were four parts to the space shuttle launch assembly—the external tank, two solid rocket boosters, and the shuttle orbiter itself. The large external tank was connected to the orbiter.

John Young and Robert Crippen ate the traditional NASA breakfast of steak and eggs on the day of their flight.

The solid rocket boosters were attached to the external tank on opposite sides. The external tank contained the liquid fuel that would be burned by the shuttle orbiter's engines. The two boosters would burn solid propellants.

Young and Crippen stepped from the van and rode the pad elevator one hundred forty-seven feet up. Technicians waiting on top escorted them across a narrow bridge to *Columbia*'s hatch. The technicians helped Young and Crippen into their positions aboard the shuttle's flight deck, then closed the hatch. Everyone cleared the pad in preparation for the launch.

This time, with only minor delays, the countdown went smoothly.

More than half a million spectators were watching the countdown from the roads and beaches and waterways surrounding Cape Canaveral. Millions more watched the countdown on live television. There was a definite air of excitement and tension as the launch drew near.

Less than a minute remained in the countdown. It looked as if *Columbia* would go this time. The launch director knew the world's first space shuttle was finally about to fly.

"Good luck, gentlemen," he said. "T minus twenty seconds and counting. T minus fifteen, fourteen, thirteen, T minus ten, nine, eight, seven, six, five, four, we have gone for main engine start."[1]

The voice could no longer be heard, as a thundering

roar echoed out across Cape Canaveral. Blinding white flames gushed from the solid rocket boosters and the orbiter's three main engines. Thick clouds of smoke and steam surrounded the pad. At three seconds past 7 A.M. eastern standard time (EST) the space shuttle *Columbia* lifted swiftly off launchpad 39A.

The public address system announcer could not contain his excitement. "We have liftoff!" he cried. "Liftoff of America's first space shuttle . . . and the shuttle has cleared the tower."[2]

The shuttle climbed higher and higher, riding a column of flame into the sky. The three main engines and solid rocket boosters were producing a combined 6.5 million pounds of thrust. The crowds cheered as the shuttle shot like an arrow toward space. It was a magnificent sight. Its velocity increased rapidly. After only two minutes *Columbia* looked like only a speck in the sky.

Columbia made its historic liftoff at three seconds past 7:00 A.M.

Communication with the shuttle was then switched over to Mission Control in Houston, Texas.

"You are Go at throttle up," said the capsule communicator, or capcom for short.

"Roger," Young said.[3]

The shuttle's main engines increased their thrust, and moments later the solid rocket boosters separated. They would parachute into the Atlantic Ocean and be recovered for reuse in a later mission. *Columbia* was now travelling at 6,200 feet per second.

The main engines burned for another six minutes. As *Columbia* shot over Bermuda, it was seventy-six miles high and moving at 13,000 feet per second.

"What a view, what a view!" Crippen said.

Anxious workers at Kennedy Space Center gathered to watch Columbia's *progress.*

"*Columbia*, you are single-engine press to MECO," said the capcom.[4] This meant that the shuttle could now get into orbit on one engine. And once in orbit the shuttle would have MECO—main engine cutoff.

Eight and a half minutes after liftoff, Young reported MECO. He and Crippen were now flying around the Earth at over 18,000 miles per hour.

The big external tank then separated from the shuttle orbiter. The tank would break up in the atmosphere and fall into the Indian Ocean. Minutes later, Young fired the orbital maneuvering system (OMS) engines for the first of two OMS burns. These OMS burns would put *Columbia* in a higher and more circular orbit. It was now flying about one hundred fifty miles above the Earth.

When the OMS burns were completed, the world's first space shuttle was successfully in orbit. The ship had performed almost perfectly since the moment it was launched.

The flight had just begun, but at NASA there was already reason to celebrate. The shuttle's launch had been almost two years behind schedule. The space program had faced tremendous challenges and solved many difficult problems designing and building the shuttle.[5] Now, after all the struggles, *Columbia* was performing better than anyone could have hoped.

John Young and Robert Crippen would spend two days in space aboard the shuttle. Everyone hoped the rest of the flight would go as smoothly as the launch.

2

Columbia in Orbit

During the launch, Robert Crippen's pulse rate had shot up to one hundred thirty beats per minute. The space rookie never denied that he was excited.

John Young, however, had been here before. Soon after the OMS burns, Young looked out *Columbia*'s windows and said, "Well, the view hasn't changed any." The pulse rate of space veteran Young had never gone above eighty-five beats per minute. He later admitted he was nervous, but "I was just so old my heart wouldn't go any faster."[1]

John Young was fifty years old. He was a serious and intelligent man who rarely showed his emotions. He had been a Navy test pilot before becoming an astronaut. In 1965, he and Gus Grissom flew in the first manned flight of the Gemini program. In July 1966, Young commanded *Gemini 10* with pilot Michael Collins.

Young was next on the crew of *Apollo 10*. This flight was the last rehearsal flight before the first Moon landing mission (*Apollo 11*). As *Apollo 10*'s command module pilot, Young orbited the Moon while Tom Stafford and Gene Cernan flew the lunar module near the Moon's surface.

Three years later, in 1972, Young commanded *Apollo 16*. He and Charles Duke spent three days on the Moon. They drove the lunar rover more than twenty miles over the Moon's surface. Young was walking on the Moon when Mission Control informed him the United States Congress had approved the space shuttle project.

"Our country needs that shuttle bad," Young said from the Moon.[2] He had no idea he would one day be its first commander.

Robert Crippen was forty-three. His friends, including John Young, called him "Crip." He was a happy and outgoing man who seemed always to be smiling. In addition to being a Navy test pilot, he had a degree in

John Young was an experienced astronaut. Columbia *was not his first voyage into space.*

13

In 1972, astronaut John Young spent three days on the Moon as part of the Apollo 16 *mission.*

aerospace engineering. Crippen became an astronaut with NASA in 1969. He had been waiting and training for his first flight into space for twelve years. It had been a long wait. But now he was at the controls of the most sophisticated space machine ever built. And he was flying it in space!

Columbia was indeed a technical marvel. The craft was designed to withstand intense temperatures, vibrations, and speeds during the mission. It could house a crew of seven to conduct experiments for more than a week in space. Its payload bay could hold up to eighteen tons of cargo. The shuttle could launch satellites, retrieve satellites, or carry a large space laboratory.

The outside of *Columbia* was covered with 60,000 individually attached heat tiles. These tiles would make the skin of the shuttle heat-resistant during its reentry into Earth's atmosphere. During the flight, the shuttle would experience temperature extremes of –250°F (–157°C) during orbit to the fiery 2,500°F (1,371°C) during reentry.

Computers aboard the shuttle made normal adjustments to the ship's orbit automatically. The shuttle's computers also continuously monitored all systems aboard the ship. This left the astronauts free to do other tasks.

The shuttle was a winged spaceship. After enduring the stresses of launch, orbit, and reentry, the spacecraft was designed to land like a plane. The landing would be a "dead-stick landing"—a landing with no power. In other words, the shuttle would not exactly *fly* to a landing, but rather *glide* to a landing.

Columbia *was the first spaceflight for astronaut Robert Crippen.*

Columbia was now in orbit, and Young and Crippen checked all its systems for the first time. When that was finished, the two astronauts crawled out of their bulky pressure suits.

Crippen got his first taste of weightlessness as he floated around the cabin. Although he had experienced no problems during his weightlessness training, Crippen did not want to take chances about getting queasy. He immediately swallowed a motion sickness pill.[3] Both the astronauts felt fine throughout the flight.

The next task was to open the payload doors.

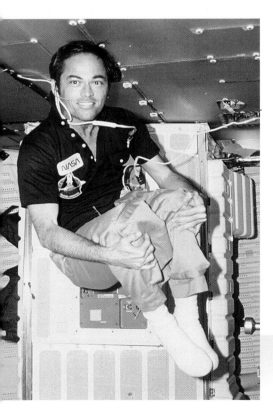

Radiators that released excess heat from the shuttle into space were on the inside of the doors. Windows at the rear of the flight deck looked out into the payload bay. Crippen opened one door and then the other.

The doors opened perfectly and the payload bay looked fine. But the two astronauts saw something else that disturbed them.

Astronaut Robert Crippen got his first taste of weightlessness aboard the space shuttle Columbia.

This photo of Columbia's *payload bay was taken by one of the astronauts as they opened and closed the payload doors. Notice the missing heat tiles on the white engine pod behind the payload bay.*

"Doors all opened hunky dory," Crippen said, pointing a television camera out the window, but "we want to show you we do have a few tiles missing."[4] Some of the heat tiles had come off!

The tiles were missing off the two engine pods that were positioned behind the payload bay. Some people were afraid that if too many of the tiles were missing, *Columbia* could not survive reentry. Crippen tried to prevent any panic.

"The starboard pod has got basically what appears to be three tiles and some smaller pieces off, and off the

Reporters questioned NASA engineers at Mission Control in Houston about the missing tiles on Columbia.

port pod looks like I see one full square and looks like a few little triangular shapes that are missing."

"Roger, Crip," the capcom said, "We see that good."

"From what we can see of both wings," Crippen continued, "it looks like they're fully intact."[5]

If the gaps they saw were the only tiles that were missing, there was no danger. The part of the shuttle that had to endure the worst heat was the underbelly. But, Young and Crippen could not see or photograph the ship's underbelly. So there was no way of knowing if tiles had come off there. Everyone simply had to hope they had not.

Reporters at Mission Control continued to question

engineers about the dangers of the missing tiles. Officials at NASA told the reporters they had no reason to be concerned about the shuttle. It had completed three orbits and was working well.

Young carried out more orbit maneuvers to boost *Columbia*'s altitude and speed. The orbit was now oval, with altitude ranging from 147 miles to 149 miles.

Crippen and Young continued to test the systems aboard the vehicle. There were a few problems, but very minor ones. Ten hours into the flight, *Columbia* and its crew were putting on a fantastic performance.

Young liked to stick to business, and he was usually not big on words. But he was impressed with the shuttle so far, and said so to Mission Control.

"The flight so far has gone as smoothly as it could possibly go," he said. "We have done every test that we are supposed to do and we are up on the time line and the vehicle has just been performing beautifully, much better than anyone ever expected it to do on the first flight, and no systems are out of shape. We owe this to a lot of people . . . they can take great pride in the ship doing so well right now."

Crippen echoed Young's comments. "It's really been super," he said. "I think we've got something that's really going to mean something to the country and the world. This vehicle has been performing like a champ."[6]

Young and Crippen had been very busy since launch, doing the various tests on the long checklist. They had little time for looking out the window at Earth or for taking pictures. Late in the day, Crippen acted as the first chef aboard the space shuttle. Being "chef" mostly meant passing out the containers of processed food placed aboard *Columbia*. They ate sandwiches and canned food specially prepared for space.

Thirteen hours after the launch it was time for the astronauts to rest. They found it difficult to sleep. Both of them were still very excited to be in space.[7]

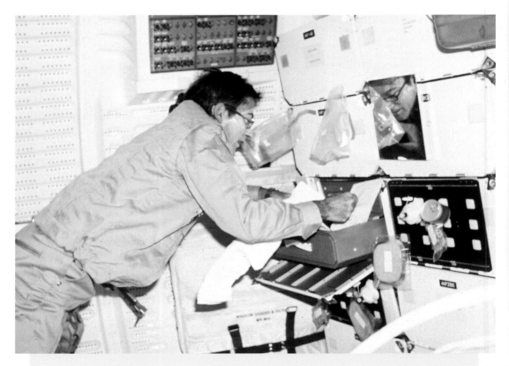

John Young cleans off his razor after shaving in space. Meanwhile, Robert Crippen acted as the first "chef" aboard Columbia. *A food tray can be seen here mounted on a locker door.*

3

Flight and Reentry

The shuttle was equipped with special sleeping hammocks. These hammocks were like sleeping bags attached to the wall, so the astronauts would not float around the cabin while asleep. But Young and Crippen did not use them. The two astronauts on this first flight had to sleep in their pilot and co-pilot seats on the flight deck. They had to be ready for any emergency. So they were not very comfortable.

After seven hours of poor sleep, Mission Control roused the astronauts awake with a specially recorded song about *Columbia*. It had a country music sound, and advised the crew to "just lay back and let 'er fly."[1]

Following breakfast, the day was filled with more tests aboard the spacecraft. Many pictures were also taken of Earth. Crippen took time out to comment on the ease of working in zero-G, that feeling of weightlessness while in orbit.

"I tell you," he said, "it's going to be tough to go back to work in the 1-G trainer after finding out how really easy it is to get around in this vehicle."

"I think you may be spoiled now," the capcom said.[2]

After lunch, the crew took a phone call from Vice President George H. W. Bush.

"Hello, Mr. Vice President," Young said, "we're just having a lot of fun up here."

"I'm glad to talk to both you and Crip," Bush said,

John Young at the controls of Columbia. *The astronauts had to be prepared for any emergency.*

"How's he behaving?" Bush knew both of them from some jogging they had done together during his visit to the Kennedy Space Center.

"I'm trying to behave pretty well," Crippen responded.

"How's it going up there? Is everything rockin' along all right?"

"The spaceship is just performing beautifully," Young said.

"Well, that's great," said Bush. "I'm sure that everyone views it as a forerunner of great things to come. I think that your trip is just going to ignite excitement and forward thinking for this country, so I really just wanted to call up and wish you the very best."[3]

After more tests of the thrusters and other systems in the evening, the second day was over. Young and Crippen slept better this time, but they were awakened by an alarm. The temperature had fallen in one of the auxiliary power units (APUs). The APUs controlled the rudder and the wing flaps, and lowered the landing gear during the landing sequence.

There were three APUs aboard. Two of them had to work in order to operate the shuttle during landing. The third was a backup. As long as the temperature in the third APU did not drop too far, they were all right. Young and Crippen monitored it. It never became a crisis.

Young and Crippen then got some more sleep. They woke at 3 A.M. EST while *Columbia* was in its thirty-first

Vice President George Bush (left) sat aboard Columbia *with John Young (center) and Robert Crippen (right) before the flight took place. During the flight, Bush spoke to the astronauts by phone.*

orbit. After breakfast they began preparing to come home. Young would now pilot *Columbia* through one of the most sophisticated reentry maneuvers ever.

He and Crippen would fire the OMS thrusters to slow the shuttle and take it out of orbit. At that time, it would be traveling at over 18,000 miles (29,000 kilometers) an hour. By the time it approached the dry lake bed at Edwards Air Force Base forty-five minutes later, it should be going no more than 250 miles (400 kilometers) an hour. In other words, it had to slow down a lot, and in a hurry.

Young and Crippen struggled back into the brown spacesuits and white helmets. All systems aboard *Columbia* were readied for reentry. The shuttle passed over California on orbit thirty-six, the last full orbit of the flight. Down below in the desert of Edwards Air Force Base, huge crowds gathered to watch the landing. They had come from all fifty states and many foreign countries to see the landing. All roads leading to the base were clogged by traffic jams six miles long.

As *Columbia* passed over Africa, Young turned the shuttle so that it was pointing tail first. "Your burn attitude looks good to us," Mission Control said.

This photo taken by the astronauts aboard the shuttle shows the great Himalayas, in India and China.

A view of the crowds gathered at Edwards Air Force Base to watch the landing of the shuttle. Traffic was backed up for six miles.

"Everything aboard looks good to us. You are Go for de-orbit burn."[4]

"We understand. Ready for de-orbit burn," Young said. Minutes later the de-orbit thrusters fired for two minutes and forty seconds. The shuttle slowed by 297 feet (91 meters) per second. When the burn was over, it had dropped from orbit and was now falling on a long arc toward California.

Young then turned the shuttle around to face nose first. In this position, the nose and belly of the shuttle faced the highest heat of reentry friction. First contact with the atmosphere began at 400,000 feet (122 kilometers). Young and Crippen were traveling at twenty-four times the speed of sound. *Columbia* would now glide through the atmosphere along a path 4,400 miles (7,100 kilometers) to California. It would be slowing down all the way.

"We're showing about eighty-five miles," Young said, telling Mission Control the shuttle's altitude.

"Roger. Moving right along," said the capcom. "Nice and easy does it, John. We're all riding with you."[5]

All communications were then lost between the space shuttle and the ground stations. The outside skin of *Columbia* heated to 2,750°F (1,510°C). The highly heated air around the shuttle became electrically

Crowds waited expectantly for the landing of the shuttle. It was a thrilling moment.

charged. This created a barrier around the shuttle that prevented any radio signals from getting through. The blackout would last about fifteen minutes.

As Mission Control waited, the ship came down to 185,000 feet (56 kilometers). It was four hundred ten miles from the landing site. Suddenly contact was reestablished.

"Hello, Houston. *Columbia* here," Young said.

"Hello *Columbia*. Houston here. How do you read?"

"Loud and clear. We're doing Mach 10.3 at 180." This meant the shuttle was traveling at more than ten times the speed of sound and was at 180,000 feet (55 kilometers).

The voices were carried over the public address system at Edwards Air Force Base. The crowds cheered at the first sound of Young's voice. Hundreds of thousands of people were pressed up to a mile-long security rope strung along the dry lake bed.

It was a thrilling moment. No spaceship had ever soared into space and landed gracefully like an airplane. This crowd would see it happen for the first time.

"We show you crossing the coast now," the capcom said. "*Columbia*, you're out of 130,000, now Mach 6.4, looking good."

Crippen was still filled with the thrill of it all. "What a way to come to California!" he exclaimed.[6]

The space shuttle *Columbia* was on its way home.

4

Welcome Home, *Columbia!*

Young and Crippen were within 150 miles (240 kilometers) of Edwards Air Force Base. As the shuttle came down, Young guided it through a series of S-turns to further slow *Columbia*'s velocity. The public address system announcer informed the crowds.

"Range is one hundred thirty miles. John Young is rolling."[1]

NASA had predicted the shuttle's speed would cause a mild sonic boom as it came near the base. *Columbia* was now near the point where it should occur.

"Seventy thousand feet, Mach 1.3. Forty-two miles," the capcom said. "*Columbia*, we show you slightly high in altitude. Coming down nicely." Young began a wide turn out high around the base. The turn would put

Columbia in position for its final approach to the landing site.

Boom! Boom!

A loud double-bang thundered out across the lake bed. The excited crowd whooped with delight as the shuttle caused not just one, but a double sonic boom.

"*Columbia,* you're subsonic now, out of 50,000 feet," the capcom said. "Really looking good. Right on the money. Turning onto final. Your winds on the surface are calm."

"That's my kind of wind," Young said.

The crowd, numbering almost a quarter of a million, then caught its first sight of *Columbia* approaching the lake bed. A fighter plane flew alongside it as it neared the landing. The shuttle was performing beautifully. Young was flying it like the pro he was.

At about 500 feet (150 meters), the landing gear came down. The shuttle was ready to come home. The crowds watched the huge space bird descend and sweep over the dusty runway. Crippen counted down the moments before landing.

"Fifty feet. Forty, thirty, twenty, ten. Five, four, three, two, one. Touchdown." *Columbia's* rear landing gear kicked up dust as the shuttle rolled along the lake bed at twice the landing speed of a normal jet airliner. Then Young gradually set the nose wheel down for a soft and perfect landing.[2]

Columbia, the first space shuttle ever, was home!

The space shuttle Columbia *makes its final approach to landing.*

The crowd at Edwards erupted in cheers, and so did the controllers in Houston.

"Welcome home, *Columbia*! Beautiful, beautiful!"

The shuttle rolled another 9,000 feet along the runway.

"Do I have to take it up to the hangar, Joe?" Young inquired playfully.

"We're going to dust it off first," the capcom said.

"This is the world's greatest flying machine, I'll tell you that," an excited Young added. "It worked super."

The shuttle came to a stop. A convoy of trucks and other service vehicles rushed up to power down the

Columbia gently touches down in the desert at the conclusion of this historic spaceflight.

spacecraft. A stair-step ramp was rolled up to the hatch. After many minutes the hatch on the side of the shuttle was opened and John Young emerged.

He made his way down the steps and handed his helmet to one of the technical crewmen. The normally serious and unemotional Young shook hands with the technical crew. He clapped like a cheerleader at the shuttle and those attending it. He marched up and down underneath *Columbia,* punching the air with excitement.[3]

The people in Mission Control smiled and laughed happily as they watched Young's display of emotion on television. They were getting a kick out of the usually quiet Young being so thrilled by the shuttle's performance. They had never seen him like that.

Minutes later Crippen came down the steps. He and Young shook hands and embraced. Then both astronauts walked around the shuttle, inspecting it. No tiles had come off the underbelly as some had feared. The ship was in beautiful shape. The first mission had gone better than anyone could have hoped.

Young and Crippen were given a quick medical checkup. Both were fine. Then they were hustled off to

A joyful John Young and Robert Crippen emerged from Columbia *after their successful return to Earth.*

meet their wives before the beginning of a short postflight ceremony.

Young told the gathering of VIPs and reporters that the flight was "a really fantastic mission from start to finish. The human race is not far from the stars."

Crippen said that his long wait to get into space was well worth it. "We are really in the space business to stay," he said.[4]

Young later tried to explain the excitement he felt at the end of the flight.

"I can't tell you what a tribute that is to the American

President Ronald Reagan (left) presents John Young and Robert Crippen with NASA's Distinguished Service Medal. Also shown are Dr. Alan Lovelace of NASA, and Vice President George Bush.

This crew patch bears the official insignia for the first flight of the space shuttle Columbia.

working man and woman," he said. "You can't imagine the variety of people who worked on this vehicle. From all walks of life, all capabilities and limitations. It's all due to their individual efforts. They proved that they can do the job. They proved it for the world to see. And I'm mighty proud to be associated with folks like that."[5]

The flight was a huge success. *Columbia* would fly many times more. The original shuttle fleet would soon expand to include *Discovery*, *Atlantis*, and *Challenger*. Both Young and Crippen commanded later shuttle flights; some experiences were good and some were bad.

Following the explosion of the *Challenger* in 1986, John Young played an important role in the work to redesign the shuttle's safety features and to prepare it for its return to space in September 1988. He serves as the associate technical director at the Johnson Space Center in Houston. He is still eligible to command future shuttle astronaut crews.

Crippen stayed on as an astronaut until he became director of the shuttle program in 1990. He served as director of the NASA Kennedy Space Center in Florida from 1992 to 1996. After serving as president of the Thiokol Propulsion Group, he retired in 2001.

The shuttle has allowed NASA to accomplish many things in space. Scientists have gained much knowledge about the Earth and the space environment. The space shuttle *Endeavor* replaced the exploded *Challenger* in 1991. Since *Endeavor,* no new shuttles have been built.

Columbia went on to complete twenty-six more successful missions from 1981 to 2002. In early 2003, the twenty-eighth *Columbia* shuttle took off from Cape Canaveral, Florida. Tragically, this mission would be its last.

The space shuttle Endeavor *replaced the exploded* Challenger *in 1991.*

5

Columbia's
Last Mission

On January 16, 2003, *Columbia* roared into space on its twenty-eighth mission. The seven astronauts on board were excited about the sixteen-day trip, which was devoted entirely to science. The astronauts were divided into two teams that would each work twelve hours a day.

Blue team members were shuttle pilot William McCool and mission specialists David Brown and Michael Anderson. The red team was made up of shuttle commander Rick Husband, mission specialists Kalpana Chawla and Laurel Clark, and payload specialist Ilan Ramon, the first Israeli astronaut. McCool, Brown, Clark, and Ramon were on their first flight into space.

After *Columbia* reached orbit, the crew floated out of their seats and started working. They had to get the twenty-foot-long SPACEHAB Research Double Module up and running. This portable pressurized laboratory, which was the size of two school buses, was located in

The STS-107 crew waves to onlookers on their way to the launchpad. Leading the way are pilot William McCool (left) and commander Rick Husband (right). In the second row are mission specialists Kalpana Chawla (left) and Laurel Clark. In the rear are payload specialist Ilan Ramon, payload commander Michael Anderson, and mission specialist David Brown.

the payload bay of the shuttle and connected to the mid-deck cabin by a tunnel. It was the first time a double module had flown, so the crew had more lab space than usual.

The extra space was needed because the astronauts were going to carry out more than eighty different studies. Most of the research taking place on this mission dealt with finding cures for diseases on Earth, such as a treatment for prostate cancer. Six of the experiments on board were from schools in Australia, China, Israel, Japan, Liechtenstein, and the United States. Students in these schools wanted to study the effects of weightlessness on bees, ants, fish, spiders, silk worms, and crystals. Can spiders spin webs in space and can ants dig tunnels?

All of the schools were part of the Space Technology

and Research Students program, or STARS. Each day the students checked www.starsacademy.com on the Internet to get updates from the astronauts. Students compared the space shuttle data with their control data at school. They had experiments in their classrooms that were exactly the same as those onboard *Columbia* except for weightlessness and radiation. So, any differences would be because of weightlessness or radiation.

The red and blue astronaut teams worked very hard to complete all of the experiments. All too soon the mission was over and it was time to return to Earth. On February 1, 2003, the five men and two women put

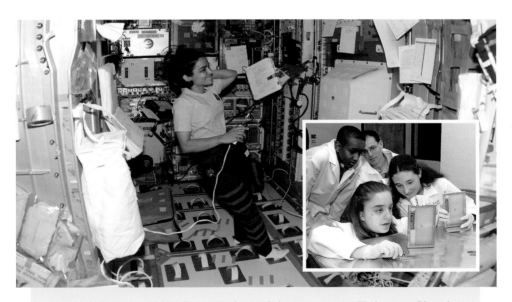

Inside the SPACEHAB research module, astronaut Kalpana Chawla checks on some experiments during the mission. Students from the STARS project (inset) check on their experiments that flew in SPACEHAB on space shuttle Columbia.

on their launch/entry suits, strapped themselves into their seats, and headed for home.

At 8:53 A.M. EST, *Columbia* flew over the California coast. It was headed for a landing at the Kennedy Space Center in Florida. A few people watching from the ground saw flashes of light around the shuttle. Astronomer Tony Beasley said, "It was like a big flare being dropped from the shuttle. It didn't seem normal."[1] As *Columbia* flew east across the United States, Mission Control in Houston watched as sensor readings blinked out. Temperature readings had risen inside the shuttle's left wing and left wheel well where the landing gear was stored.

Just before 9:00 A.M. EST, Mission Control radioed, "*Columbia*, Houston, we see your tire-pressure messages and we did not copy your last."[2] Mission Commander Rick Husband replied "Roger, uh . . ." and then there was silence and static. The orbiter was sixteen minutes from landing. At Kennedy Space Center in Florida, the astronauts' families and friends peered into the sky. Ilan Ramon's wife said, "The clock ticked and we counted, and it was quiet when we should have heard a noise. There were supposed to be sonic booms, but they didn't come."[3]

Tragically, while traveling 12,825 miles per hour and at an altitude of 38 miles, the shuttle broke apart over Texas. "It was like a car hitting the house or an explosion. It shook that much," said John Ferolito of Carrolton, Texas.[4]

Scientists do not know exactly what went wrong. They do know, however, that in order to get into space, a shuttle must go really fast. To come back, it has to slow down. This is done with braking jets and friction from entering Earth's atmosphere. The friction causes extreme heating. The shuttle's tiles are a shield against this heat, so that the heat does not melt equipment, such as tires, that are tucked inside the wings while in space. Somehow, the shield was penetrated and hot gas got inside the left wing, near the tires. The heat melted wires that control the shuttle's movement. The shuttle's computers tried to keep the shuttle at the right angle during reentry, but could not. As a result, the shuttle broke apart.

Debris from the space shuttle Columbia *streaks across the sky over Tyler, Texas. The shuttle was scheduled for a morning landing at the Kennedy Space Center in Florida.*

Debris rained from the sky. Parts of the shuttle were scattered over more than three hundred miles and several states, mostly Texas. Some landed in the woods or in the Toledo Bend Reservoir, a huge lake of water. Local residents were warned not to touch any of the pieces. Some debris was dangerous because it had rocket fuel in or on it that might explode or make toxic fumes.

Search teams made up of policemen, firemen, volunteers, and National Guard troops fanned out into the countryside. They fought their way through dense underbrush, looking for pieces of *Columbia* and remains of its crew. The nose cone was found and hauled out of the woods on a tractor. Parts of the crew cabin and a two-foot section of wing were also located. All of the debris was taken to the Kennedy Space Center in Florida. Engineers and scientists laid the debris out by location. Remains of all seven astronauts were taken to Dover Air Force Base in Delaware and identified. Then they were released to their families for burial.

A memorial service was held at NASA's Johnson Space Center in Houston, Texas. More than ten thousand aerospace workers attended, along with the friends and families of the victims. President George W. Bush told the grieving crowd, "Our nation shares in your sorrow and in your pride. We remember not only one moment of tragedy, but seven lives of great purpose and achievement."[5]

A *Columbia* Accident Investigation Board was formed

The debris from the Columbia *shuttle is being shipped to a hangar at the Kennedy Space Center.*

to learn what caused the tragedy. Something happened to the left wing of *Columbia*, but what? Search parties continued to comb the countryside, looking for pieces of *Columbia* that could provide answers. Dive teams searched the Toledo Bend Reservoir while helicopters flew over the Sabine National Forest, looking for debris. Scientists will continue to work hard to determine the cause of the *Columbia* tragedy, but they may never know why the mighty shuttle exploded into pieces in the morning sky.

The families of the seven astronauts lost aboard *Columbia* thanked NASA and people around the world for their "incredible outpouring of love and support." They summarized their feelings about the future of spaceflight: "Although we grieve deeply, . . . the bold exploration of space must go on."[6]

CHAPTER NOTES

Chapter 1. A New Kind of Spaceship

1. Richard S. Lewis, *Voyages of Columbia* (New York: Columbia University Press, 1984), pp. 128–129.

2. Ibid., p. 129.

3. Ibid.

4. Ibid., p. 130.

5. John and Nancy Deward, *History of NASA* (Greenwich, Conn.: Brompton Books, 1984), pp. 174–176.

Chapter 2. *Columbia* in Orbit

1. Peter Bond, *Heroes In Space: From Gagarin to Challenger* (New York: Basil Blackwell, Inc., 1987), p. 399.

2. "Spaceflight Part 4: The Road Ahead," narrated by Martin Sheen, PBS Video (1985).

3. Bond, p. 400.

4. Richard S. Lewis, *Voyages of Columbia* (New York: Columbia University Press, 1984), p. 134.

5. NASA, *First Flight of the Shuttle Columbia* (Houston, Tex.: Finley-Holiday Film Corp., 1981).

6. Ibid.

7. Lewis, p. 136.

Chapter 3. Flight and Reentry

1. Richard S. Lewis, *Voyages of Columbia* (New York: Columbia University Press, 1984), p. 136.

2. Ibid., p. 137.

3. NASA, *First Flight of the Shuttle Columbia* (Houston, Tex.: Finley-Holiday Film Corp., 1981).

4. Lewis, p. 140.

5. NASA, *First Flight of the Shuttle Columbia.*

6. Peter Bond, *Heroes in Space: From Gagarin to Challenger* (New York: Basil Blackwell, Inc., 1987), p. 401.

Chapter 4. Welcome Home, *Columbia!*

1. Richard S. Lewis, *Voyages of Columbia* (New York: Columbia University Press, 1984), p. 141.

2. NASA, *First Flight of the Shuttle Columbia* (Houston, Tex.: Finley-Holiday Film Corp., 1981).

3. Michael Collins, *Liftoff!* (New York: Grove Press, 1988), p. 216.

4. Peter Bond, *Heroes In Space: From Gagarin to Challenger* (New York: Basil Blackwell, Inc., 1987), pp. 402–403.

5. NASA, *First Flight of the Shuttle Columbia.*

Chapter 5. *Columbia*'s Last Mission

1. Thomas, Evan. "Out of the Blue," *Newsweek*, February 10, 2003, p. 24.

2. Ibid., p. 28.

3. "Farewell Columbia," *People*, February 17, 2003, p. 95.

4. Associated Press, February 1, 2003.

5. "Columbia's Last Mission," *Houston Chronicle*, February 5, 2003, p. A 24.

6. CNN.com, "Astronauts' families: Space exploration 'must go on'," February 3, 2003, <http://www.cnn.com/2003/US/02/03/sprj.colu.family.statement/index.html> (February 22, 2003).

GLOSSARY

auxiliary power units (APUs)—The additional or reserve electrical power sources aboard the space shuttle.

external tank—The large tank attached to the shuttle that supplies liquid fuel to the shuttle orbiter's main engines during liftoff.

payload bay—The large area of the shuttle orbiter used to carry cargo into space.

reentry—The return through the Earth's atmosphere.

satellite—An artificial object put into orbit.

shuttle orbiter—The reusable delta-winged spacecraft that, with the solid rocket boosters and external tank, makes up the space shuttle or space transport system.

solid rocket booster—A rocket that uses explosive powders for fuel. This type of booster helps launch the space shuttle off the pad.

sonic boom—The explosive sound created when the shock wave formed at the front of an aircraft traveling at supersonic speeds reaches the ground.

space shuttle—The reusable U.S. space transportation system that carries astronauts, satellites, and experimental scientific equipment into orbit around the Earth.

zero-G—The sensation of weightlessness, or zero gravity, that is experienced by astronauts in space.

FURTHER READING

Books

Fichter, George S. *Space Shuttle*. New York: Franklin Watts, 1990.

Gold, Susan D. *To Space & Back: The Story of the Shuttle*. New York: Crestwood House, 1992.

Lassieur, Allison. *The Space Shuttle*. Danbury, Conn.: Children's Press, 2000.

Maynard, Christopher. *The Space Shuttle*. New York: Larousse Kingfisher Chambers, 1994.

Reichhardt, Tony, ed. *Space Shuttle: The First 20 Years—The Astronauts' Experiences in Their Own Words*. New York: DK Publishing, 2002.

Spangenburg, Ray, Kit Moser and Diane Moser. *Onboard the Space Shuttle*. Danbury, Conn.: Franklin Watts, 2002.

Internet Addresses

To see data from the student experiments aboard *Columbia*:
<http://www.starsacademy.com/sts107/index.html>

For more information about NASA and space travel:
<http://spacelink.msfc.nasa.gov>

To learn more about human spaceflight:
<http://spaceflight. nasa.gov>

INDEX